BEI GRIN MACHT SICH IHR WISSEN BEZAHLT

AF135625

- Wir veröffentlichen Ihre Hausarbeit,
  Bachelor- und Masterarbeit

- Ihr eigenes eBook und Buch -
  weltweit in allen wichtigen Shops

- Verdienen Sie an jedem Verkauf

Jetzt bei www.GRIN.com hochladen
und kostenlos publizieren

Safety Measures in Autonomous Vehicles Technology

Bishnu Sharma

Bibliografische Information der Deutschen Nationalbibliothek:

Die Deutsche Nationalbibliothek verzeichnet diese Publikation in der Deutschen Nationalbibliografie; detaillierte bibliografische Daten sind im Internet über http://dnb.d-nb.de abrufbar.

ISBN: 9783346704863
Dieses Buch ist auch als E-Book erhältlich.

© GRIN Publishing GmbH
Nymphenburger Straße 86
80636 München

Alle Rechte vorbehalten

Druck und Bindung: Books on Demand GmbH, Norderstedt Germany
Gedruckt auf säurefreiem Papier aus verantwortungsvollen Quellen

Das vorliegende Werk wurde sorgfältig erarbeitet. Dennoch übernehmen Autoren und Verlag für die Richtigkeit von Angaben, Hinweisen, Links und Ratschlägen sowie eventuelle Druckfehler keine Haftung.

Das Buch bei GRIN: https://www.grin.com/document/1262757

FACTORS INFLUENCING THE SAFETY MEASURES IN AUTONOMOUS VEHICLES TECHNOLOGY

Bishnu Sharma
M.Sc. Software Engineering
University of Europe for Applied Sciences, Potsdam,
Germany

Abstract

Safety and protection measures are very crucial when it comes to autonomous driving. Safety is a condition of being protected from unlikely to cause danger and a requirement or key challenge to autonomous vehicle. Any accident, engine or sensors failure may result in severe injury or loss of life. In this report, we are presenting the factors that influence the safety measures of autonomous vehicles and evaluate the performance of some of them such as Lidar sensors, Rider sensors, Actuators, Complex algorithms by using On-road testing. Also, different driving automation levels are explained that should be taken while designing autonomous vehicle safety.

Keywords: Safety measure, Automation, Sensors, Cyber Security, Perception

Introduction

An autonomous vehicle (AV) or driverless vehicle (DV) is one that can operate and execute required functions without human involvement by sensing its surroundings. In addition, AV employs a completely automated driving system to allow the vehicle to adapt to external conditions that a human driver would handle. The major transportation capabilities of a typical car may be fulfilled by an autonomous vehicle. The key distinction is the Driving Automation System (DAS) developed for self-driving vehicles. DAS adds driving automation to the vehicle platform, allowing for major changes in transportation that minimize collisions, energy consumption, pollution, and congestion costs. The SAE prefers the word automated to autonomous. One reason is because the term autonomy has meanings other than electromechanical. A completely self-driving automobile would be self-aware and capable of making its own decisions. To execute software,

1

autonomous vehicles rely on sensors, actuators, complicated algorithms, machine learning systems, and powerful computers. AV is a life-saving system. Any AV malfunction might result in serious human injury or even death. The goal of safety is to safeguard the system against unintentional breakdowns in order to avoid dangers. Autonomous vehicle safety encompasses mechanical system safety as well as electrical and electronic system safety. In this paper, we have presented a factor that influences the safety measures in autonomous vehicles throughout the early development phases by synchronizing the safety lifecycle, which is based on the SAE J3061, SAE J3016, and ISO 26262 standards.

The article's goals are thus to carefully investigate the safety risks associated with autonomous technology applied to vehicle applications. In section 2: the levels of automation in cars are discussed, in section 3: the sorts of errors and their probable causes are thoroughly examined, section 4: discussed about the tools used in Avs, Section 5: investigates the present state of on-road testing and accidents, and section 6: discusses the opportunities and difficulties for AV safety research.

2.0 Levels of Automation

The definition of AVs is critical for regulation makers who want to limit the impact of this technology on conventional road users including other vehicles, pedestrians, bicycles, and even construction workers. As previously said, the amount of vehicle automation is determined by the complexity of the autonomous technology used, the perceptual range of the environment, and the degree to which a human driver or vehicle system is involved in the driving choice, which is strongly tied to AV safety. This section summarizes and compares the definitions of automation levels from various organizations. Sheridan and Verplank first described the standard definition of automation levels in 1987, and it was later modified by Parasuraman Tal. in 2000. The responsibilities of the human operator and the vehicle technology in the driving process are used to establish ten degrees of automation. Level 1 indicates there is no automation, and all decisions and actions are made by humans. Levels 2–4 systems can recommend a whole set of alternative decision or action plans, but human supervisors determine whether or not to execute the suggested activities. From Level 5, the system is capable of carrying out a decision with the consent of a human operator. At Level 6, the technology permits the human driver to

2

respond within a specific time frame prior to the autonomous action. At Level 7, following an automatic action, the system will notify the human supervisor, whereas at Level 8, the system will notify the human supervisor. The system will not notify the human supervisor unless specifically requested. At Level 9, the system will decide whether or not to notify the human supervisor following an automated action. Level 10 denotes total automation, with no human intervention.

The stages of automation in aerospace engineering vary, but six levels are commonly established, which is known as the Pilot Authorization and Control of Tasks (PACT) framework.

Levels 0 through 5 are assigned to this automation system. Level 0 indicates that there is no computer autonomy, while Level 5 indicates that the systems can be fully automated but can still be halted by a human pilot. The PACT proposed four aided modes in addition to human commanded and autonomous modes, based on the operational relationships between human pilots and the equipment. The details of the six levels can be found in The National Highway Traffic Safety Administration (NHTSA) of the United States recognized five stages of automation in automobile engineering. The automation levels in this system are separated into five categories, numbered from 0 to 4. Level 0 symbolizes total driver control of the cars, with no automation. Level 4 symbolizes fully self-driving automation, in which the vehicle can monitor external circumstances and conduct all driving responsibilities. As can be observed, the majority of present autonomous vehicle development operations may be classified as Level 3, limited self-driving automation, where drivers can take over driving in specific situations. Recently, The National Highway Traffic Safety Administration (NHTSA) has accepted a more generally used definition of AVs based on the Society of Automotive Engineers (SAE) [2], which is constantly revised. SAE defines six degrees of vehicle automation, ranging from 0 (no automation) to 5 (complete driving automation), based on how much the human aspect is necessary for the automation system. SAE defines six degrees of driving automation, which are frequently used by automotive manufacturers, regulators, and policymakers [2]. These degrees of automation are classified based on the involvement of the human driver and the automation system in controlling the following driving tasks: (i) execution of steering and throttle control, (ii) monitoring the driving environment, (iii) dynamic driving task (DDT) fallback, and (iv) system capabilities of several autonomous driving modes. Levels 0-2 rely on the human driver to conduct part or all of the DDT, whereas Levels 3-5 reflect conditional, high, and complete driving automation, respectively, implying that the system can do all of the DDT when engaged. This detailed characterization of vehicle automation levels is commonly utilized in contemporary AV development operations. The following are the six degrees of driving automation specified by the Society of Automotive Engineers (SAE) [2]:

(i) **Level 0 (No Automation):** All driving tasks are accomplished by the human operator.

(ii) **Level 1 (Driver Assistance):** The car is controlled by a human, although driving is aided by an automated system.

(iii) **Level 2 (Partial Driving Automation):** In the vehicle, combined automated features are used, but the human operator still observes the surroundings and supervises the driving process.

(iv) **Level 3 (Conditional Driving Automation):** The human operator must be ready to operate the vehicle at any time.

(v) **Level 4 (High Driving Automation):** Under certain situations, the automation system can drive itself, and the human driver may be able to operate the vehicle.

(vi) **Level 5 (Full Driving Automation):** Under any situations, the automation system is capable of driving autonomously, and the human driver may be able to operate the car.

The diverse definitions of automation levels by different organizations show that human operators and vehicle systems can be involved in driving activities to varying degrees. As a result, the safety considerations for somewhat, highly, and totally autonomous vehicles might differ significantly. When AVs are operated in no automation, partial automation, or high automation modes, the interaction between human operators and machines can pose a significant challenge for AV safety, when AVs are operated in fully automated modes, the liability of the software and hardware becomes a critical issue. In other words, as more autonomous technology is used in vehicles, the complexity of the autonomous system increases, posing difficulties to system stability, dependability, and safety. As a result,

4

theoretical study of probable AV errors will be critical in understanding the existing AV safety status and forecasting future safety levels.

3.0 Types of Error for Autonomous Vehicles

Different forms of mistakes may be developed when more autonomous procedures are used. If such mistakes are not managed appropriately, they might lead to significant safety hazards. A comprehensive investigation of various forms of mistakes or accidents in AV technology will aid in understanding the present state of AV safety. It should be emphasized that the number of accidents documented in the literature for AVs is quite low when compared to regular cars. However, this does not imply that present AVs are inherently safer than human-controlled cars. Because AV technology is still in its early stages of commercialization and is still a long way from completely autonomous driving, additional road testing and accident databases should be conducted. The dependability of the AV architecture and its accompanying hardware and software determines AV safety. However, because AV design is so reliant on the amount of automation, AV safety may exhibit distinct patterns at different phases. Even at the same automation level, the architecture of AVs might differ between studies.

The overall design and key components of AVs are depicted in Figure 1. A typical AV is made up of a sensor-based perception system, an algorithm-based decision system, an actuator-based actuation system, and system interconnections. Ideally, all AV components should work well so that AV safety may be assured.

3.1. *Accidents Caused by Autonomous Vehicle*: AV safety concerns or accidents are closely connected to AV faults performed at various automation levels. In general, such mistakes may be classified using the aforementioned architecture. (Figure:1).

3.1.1. *Perception Error*: The perception layer is in charge of collecting data from many sensing devices in order to perceive environmental circumstances in real time. The intricacy, dependability, and applicability of Avs are the primary determinants of their evolution. as well as the maturity of sensor technologies. Light detection and ranging (LIDAR) sensors, cameras, radars, ultrasonic sensors, touch sensors, and global positioning systems are examples of sensors used for environmental perception (GPS). The role and capability of different sensing technologies may be discovered elsewhere. It should be highlighted that any inaccuracies in perception of other road users' status, position, and movement, traffic signals, and other dangers may cause safety problems for AVs. Figure 2 highlights the history and probable future progress of AV technology depending on the unique sensor technology used in vehicle systems, with data from. Proprioceptive sensors, such as wheel sensors and inertial sensors, were introduced around the close of the twentieth century. and odometry are often used in-vehicle systems to perform the functions of traction control system, antilock braking system, electronic stability control, antiskid control, and electronic stability program. Many attempts were made in the first decade of the twenty-first century to provide information, warning, and comfort while driving using exteroceptive sensors such as sonar, radar, lidar, vision sensors, infrared sensors, and global navigation satellite systems. Navigation, parking assistance, adaptive cruise control, lane departure warning, and night vision are all available on the cars.

Sensor networks deployed in both vehicle and road systems have been incorporated in the current transportation system during the last decade for the goal of automated and cooperative driving. Advanced autonomous features, such as collision prevention and mitigation, and automated driving, will be enabled, and drivers will finally be liberated from the driving process. The perceived data may also originate through the communication between the AVs and the associated infrastructure other cars, the Internet, and the cloud, depending on the level of vehicle automation. The three primary sources of perception mistakes are hardware, software, and communication. Because the perception system is primarily reliant on sensing technologies, perception mistakes may arise from the hardware, especially sensors. For example, sensor deterioration and failure may result in server perception mistakes, confusion of the decision system, and risky driving behaviors. As a result, dependable and fault-tolerant sensing technology might

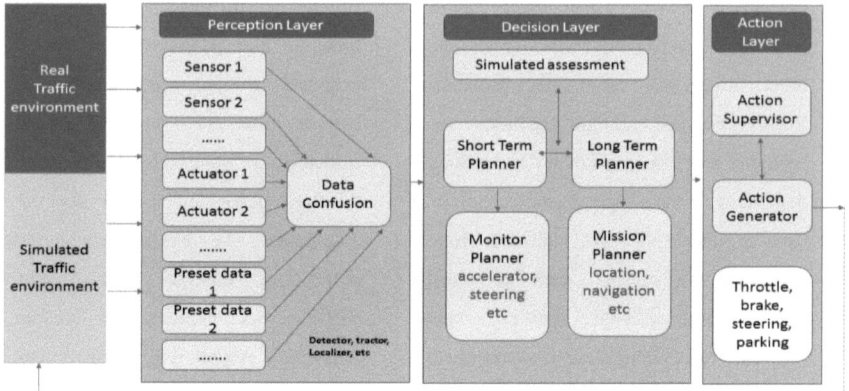

Figure1: A typical autonomous vehicle system architecture

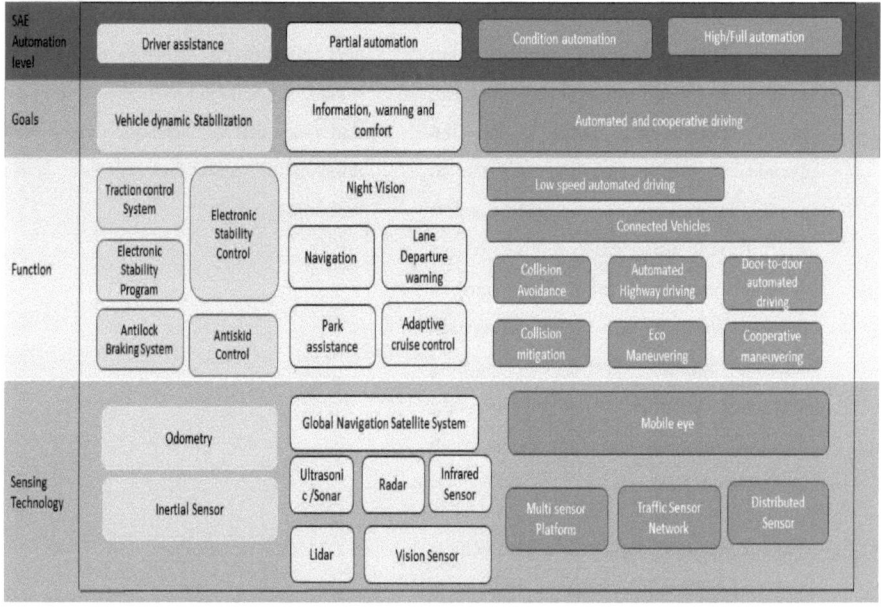

Figure 2: Past and potential future evolution of autonomous vehicle technology.

be a viable solution to such problems. Furthermore, perception mistakes may come from software failure, and this form of error would create misleading to the decision and action layers, potentially failing the mission objectives or causing

7

safety issues. When the AVs reach complete automation, communication failures will become critical. Communication faults can occur as a result of errors in communication between AVs and relevant infrastructure, other road users, and the Internet . Interpersonal contact is essential in today's transportation system. Road users, such as cars, pedestrians, bikers, and construction workers, interact with one another to coordinate movements and guarantee road safety, which are key needs for AVs. Gestures, facial expressions, and vehicular gadgets are examples of communication techniques, and the comprehension of these signals may be altered by a range of elements like as culture, environment, and experience, which are also major problems for AV technology.

3.1.2. Decision Error: The decision layer evaluates all of the perception layer's processed input, makes judgments, and provides the information required for the action layer. Situational awareness feeds the decision-making mechanism for both short-term and long-term planning. Short-term planning entails trajectory generation, obstacle avoidance, and event and maneuver management, whereas long-term planning entails mission planning and route planning. The system or human factors are mostly responsible for decision mistakes. A good AV system will only take over driving or alert drivers, when necessary, with a low false alarm rate but adequate positive performance (e.g., safety level). As AV technology advances, the false alarm rate can be lowered significantly while maintaining adequate precision to fulfill safety criteria. However, if the algorithm is unable to recognize all dangers effectively and efficiently, AV safety would be jeopardized. It should be noted that it may take a few seconds for drivers who are preoccupied with other chores to respond and take over control from the automated vehicle, which introduces uncertainty into safe AV operation. Unfortunately, AV technology is not yet totally dependable; so, when the AV system fails or is restricted in performance capabilities, the human driver must take over the driving process, overseeing and monitoring the driving responsibilities. As a result of the shifting role of a human driver in AV driving, inattention, lower situational awareness, and manual skill deterioration may occur . As a result, while building AVs from a human-centered perspective, how to safely and effectively re-engage the driver when the autonomous systems fail should be considered.

3.1.3. Action Error: Following the decision layer's order, the action controller will operate the steering wheel, throttle, or brake for a typical engine to change direction and accelerate or decelerate. Furthermore, the actuators monitor the feedback variables, and the feedback data is used to produce new actuation decisions.

Action mistakes caused by actuator failure or breakdown of the power train, controlling system, heat management system, or exhaust system, similar to traditional drive systems, can lead to safety issues. A human driver, on the other hand, would be able to detect such safety hazards while driving and pull over in a timely manner. The complete automated driving system would have difficulties in determining how the car learns in these instances and responds to these low-frequency but deadly faults of important vehicular components. As a result, classic automobile accident reconstruction would be necessary.

4.0 Tools used in safety Measure for Autonomous Vehicles

Sensors, actuators, complicated algorithms, ML frameworks, and cutting-edge computers are used to perform programming in autonomous vehicles.

Based on a variety of sensors installed in various parts of the vehicle, AV create and maintain a map of their environmental elements. Radar sensors monitor the status of nearby cars. Video cameras detect traffic signals, read street signs, monitor various cars, and look for pedestrians. Lidar (light

detection and ranging) sensors measure distances, differentiate street boundaries, and recognize path markers by skipping light pulses off the vehicle's ambient elements. When parking, ultrasonic sensors

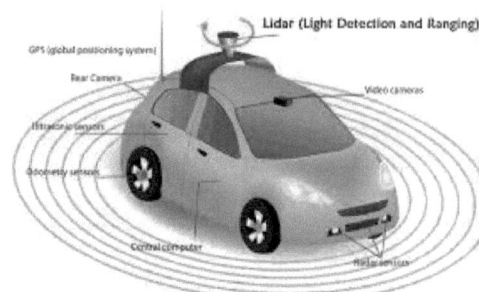

in the wheels distinguish obstacles and various cars.

When we look on LiDAR technology its by making the car drive itself with advanced sensors that can avoid a crash before it happens.

By a light sensing technology that creates a 3D map of a car's surroundings using a Laster and receiver.

Advantage of LiDAR Accuracy

Above all, LiDAR technology offers incredibly accurate, consistent results. The short wavelength can even detect small objects and create exact 3D models, making it possible to determine what the objects are, whether it's a tree, person, or wall.

Speed

The sensor sends out laser pulses and receives them back in nanoseconds, making it possible to scan large areas in a fairly short period of time and still get a high volume of data can be collected from a variety of places that are inaccessible, such as high mountains, dense forests and hard to reach areas can be easily mapped with LiDAR technology.

Automated Functionality

LiDAR technology consists of primarily automated processes, and while experienced pilots are necessary to operate the equipment, it's more efficient than other methods of surveying that require more hands-on involvement.

Low Cost

Given the speed and large area that can be scanned coupled with the highly accurate results, LiDAR is significantly less expensive than other methods of land surveying and mapping. It is an affordable way to produce complex topographical surveys.

Disadvantages of Using LiDAR

Speed, cost, and the absolute volume of highly accurate data tend to make LiDAR the right option, but it's important to know the disadvantages as you plan it requires experience to operate.

It takes previous surveying experience to take check shots, run base stations, and check-in to benchmarks. The LiDAR is complex in nature and requires a deep understanding of the sensor. Purchasing High-End LiDAR sensors are costly. If you are trying to set up your own LiDAR shop it requires heavy investment into the LiDAR sensor and the personnel.

Some companies are against LiDAR claiming that they are expensive sensors and unnecessary technology since humans can drive under all conditions by using only vision.

However, we can't say at the moment which side are right or wrong for different reasons, while we are still lacking enough reaches on the ground studies, since most of the studies are done in good conditions and they still can't function as expected bad weather conditions.

5. On-Road Testing and Reported Accidents

This section analyzes publicly accessible data for on-road AV testing,

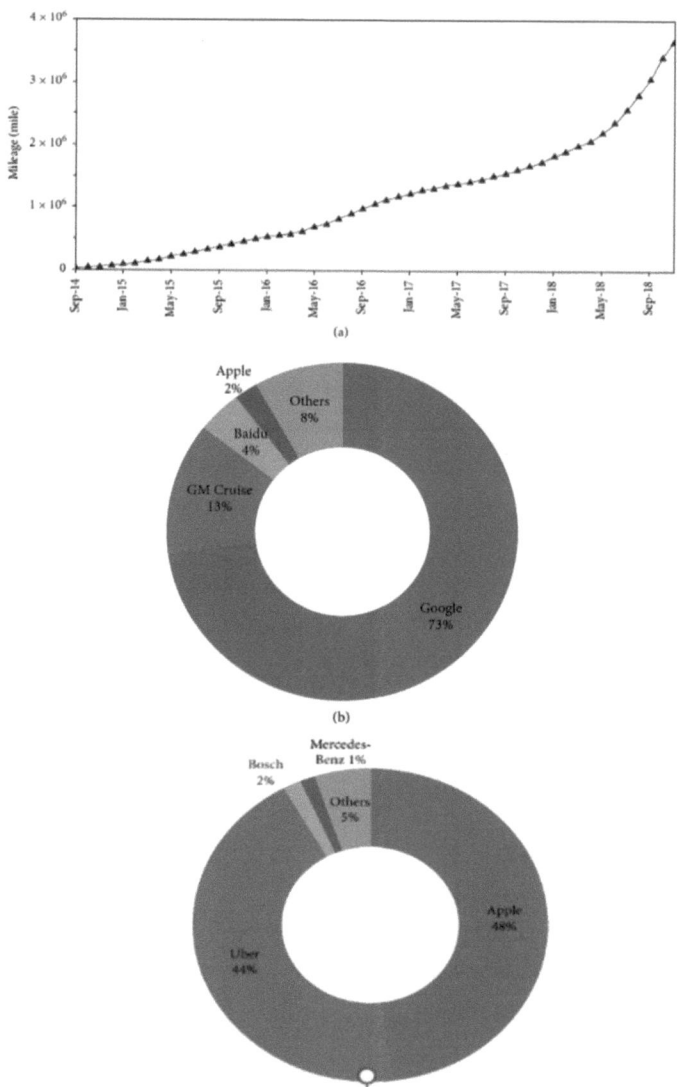

Figure 3: (a) Cumulative mileage, (b) breakdown of mileage, and (c) breakdown of disengagements based on various manufacturers. (Data are statistically analyzed from the reports by the State of California Department of Motor Vehicles between September 2014 and November 2018; the data from Waymo and Google are combined and noted as Google in this figure).

including disengagement and accident reports, to provide a firsthand knowledge of AV safety. This section investigates two common data sources from the California Department of Motor

Vehicles (USA) and the Beijing Innovation Center for Mobility Intelligent (China).

(5.1) California Department of Motor Vehicles: The State of California

reported by the State of California Department of Motor Vehicles as of April 2019, with 621 disengagement reports statistically examined between 2014 and

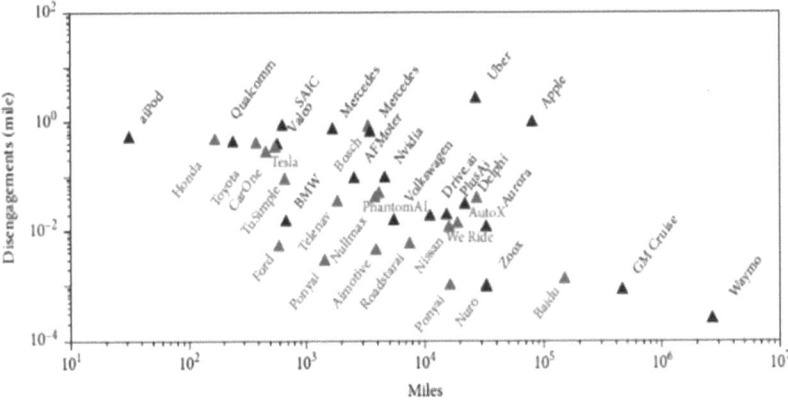

FIGURE 4: Number of disengagements vs. autonomous miles according to reported data from various manufacturers.

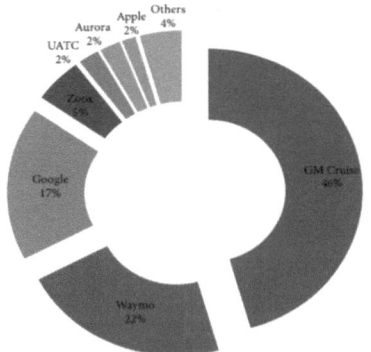

FIGURE 5: Breakdown of autonomous vehicle accident reports. (Data are statistically analyzed from the reports by the State of California Department of Motor Vehicles between September 2014 and November 2018).

Department of Motor Vehicles reports on safety concerns disclosed during on-road testing, as reflected by measurements and actual accidents. This section examines the disengagement and accident reports

2018. Figure 3 depicts the statistical situation. Manual takeovers were documented when AV operators decided to manage the vehicles manually rather than using automated systems when it was deemed essential. These incidents can be triggered by challenging actual driving circumstances, such as emergency vehicles, construction zones, or unexpected items on the road. The recognition of a problem with perception, motion planning, controls, or communications can result in software disengagement. For example, if the sensors are unable to detect and track an item in the surrounding environment, human drivers will assume control of the vehicle. A disengagement event will occur if the decision layer fails to provide a motion plan and the actuator responds late or incorrectly.

Figure 5 depicts the split of real AV incident reports from the California Department of Motor Vehicles between 2014 and 2018. After statistically analyzing 128 accident reports, the top four reporters are GM Cruise (46 %), Waymo (22 %), Google (17 %), and Zoox (5 %). It is worth noting that Waymo evolved from the Google Self-Driving Car Project in 2009 [80]. In the last four years, 36.7% of the accidents occurred in traditional manual-control mode, while the remaining 63.3% occurred in autonomous driving mode. This suggests that autonomous technology is still in requires more intensive on-road testing before it can be completely applied to the AVs. It is also noteworthy to note that AVs cause just a small part (approximately

6.3 %) of overall accidents, whereas 93.7 % of accidents are caused by humans, by other parties, such as walkers, cyclists, and motorcycles as well as conventional vehicles. This denotes that a more research into the prospective AV operating approach to reduce passive accidents might significantly enhance AV safety.

Figure 6 shows the relationship between reportable accidents and the overall distance for the California-tested Avs prior to 2017, it can be shown that the number of reportable

accidents increase slowly with overall testing kilometers a rate of 1.7105 accidents per mile yet, from 2017 and in 2018, the rate of growth is 4.9105 accidents per mile that is in California, immature technology is being used to recently tested AVs, as well as an increasing number of AVs being evaluated concurrently.

5.2. Beijing Innovation center for Mobility Intelligent: The Beijing Innovation Center for Intelligent Mobility has revealed on-road AV testing in restricted metropolitan locations in 2018 [27]. Since March, the autonomous driving mileage has reached 153,565 km (equivalent to 95,420 miles) at the end of December 2018 (see Figure 7(a)). Baidu (90.8 %), Pony.ai (5.6 %), NIO (2.8 %), and Daimler AG are the top four manufacturers (0.5 %). However, no disengagement or accident reports are currently available. It would be useful if accident-related information could be made public, and the shared knowledge might help all manufacturers encourage the use of autonomous vehicles technology and build the customers confidence in AVs.

6. Opportunities and Challenges

AV technology will help our community from a variety of aspects, including improved transportation safety, reduced traffic congestion, the removal of people from the driving process, and an economic and environmental effect. As a result, sophisticated AV technology has piqued the interest of both academics and industry, indicating a wide range of potential for AV advancement. However, before AVs can be pushed in markets, significant testing is required, and new difficulties from the adopted software, hardware, vehicle system, infrastructure, and other road users must be handled.

6.1. Opportunities: One argument in favor of AV technology development is that it will eliminate many traditional work prospects. However, as technology advances, more occupations will be produced in reality. AV development requires extensive on testing software, hardware, vehicle system, sensing devices, communication systems and other

components. Thanks to AV technology that human operators can be relieved of the driving process, and time can be better controlled. People will work, play, and learn more effectively as AV technology is promoted. Furthermore, the prevailing way of life would be disrupted. For example, the methods of driving instruction and the driver's license test would be altered. In other words, not only the AV sector but also the non-AV industry may benefit.

Traditional modes of transportation may potentially be altered by AV methods. The need to relieve vehicle operators of the burden of driving has fueled the development of intelligent vehicle grids, which employ a platform of sensors to collect data from the surrounding environment, including other road users and road signs. These signals will be provided to the drivers and infrastructures to enable safe navigation, pollution reduction, improve fuel efficiency, and to control traffic efficiently. Stern et al. conducted a ring road experiment with both autonomous and human-operated vehicles, and their findings show that a single AV can control the traffic flow of at least 20 human-operated vehicles, with significant improvements in vehicle velocity standard deviation, excessive braking, and fuel economy. Liu and Song looked at two types of AV lanes: dedicated AV lanes and

AV/toll lanes. The dedicated AV lane exclusively enables AVs to pass, but the AV/toll lane lets human-operated cars to pass by paying additional tolls, and their modeling findings show that using both techniques improve system performance [86]. Gerla et al. reviewed the Internet of Vehicles capable of communications, storage, intelligence, and self-learning [60]. Their work indicated that the communication between automobiles and the Internet will drastically alter public transportation, making it more efficient and environmentally friendly. As a result, existing transportation system must be updated for AVs.

Driving simulators have received a lot of attention for its ability to simulate automated driving conditions and accident situations in a virtual reality environment. Driving simulators allow for the efficient study of driving behaviors, take-over requests, car-following maneuvers, and other human aspects. This can reduce the risk of putting drivers in risky situations and simulating the decision-making process and the consequences.

6.2. Difficulties: Due to safety concerns, the widespread use of AVs remains a challenge. The AVs will be promoted if the following issues are addressed:

6.2.1. Reducing Perception Errors: To reduce perceptual errors, it will be difficult to identify, localize, and categorize items in the surrounding environment. Furthermore, recognition and comprehension of human characteristics like as posture, speech, and mobility will be critical for AV safety.

6.2.2. Reducing Decision Errors: A dependable, robust, and efficient decision-making system should be designed in order to respond accurately and quickly to the ambient environment. This should be accomplished through thorough and stringent hardware and software testing. Furthermore, making right judgments in complex settings remains tough, for example, what should be the decision if the AVs will have to hurt pedestrians to avoid fatal accidents due to sudden system faults or mechanical failures.

6.2.3. Reducing Action Errors: To accomplish safe AVs, actuators must be able to connect with decision systems and execute orders from either human operators or automated systems with great dependability and stability.

6.2.4. Cybersecurity: As autonomous technology advances, AVs will need to electronically connect with roadside infrastructure, satellites, and other cars (e.g., vehicular cloud). One of the most pressing challenges for AVs will be how to ensure cyber-security.

6.2.5. Interaction with the Traditional Transportation System: In metropolitan locations, AVs and conventional cars will share public roadways, and interactions between AVs and other road users, including traditional vehicles and pedestrians, will be difficult. It is difficult for other road users to distinguish the sorts of vehicles that they are interacting with. For pedestrians, this uncertainty may lead to stress and altered crossing decisions, especially when the AV driver is preoccupied with other activities and does not make eye contact with them. The work of Rodr'guez Palmeiro et al. recommended that fine-grained behavioral measures, such as eye-tracking, can be examined further to identify how pedestrians respond to autonomous vehicles [4].

6.2.6. Customer Acceptance: The key challenges restricting the commercialization of AVs are safety, cost [17], and public interests, with safety being the most important concern that may dramatically impact public perception of new AV technology.

7. Summary and Concluding Remarks

Fully autonomous vehicles (AVs) will allow cars to be driven totally by automated systems, allowing human operators to be engaged in things other than driving. Although AV technology will benefit both people and communities, safety concerns remain technological barriers to the effective commercialization of AVs. The levels of automation established by several organizations in various industries are collected and contrasted in this review article. The Society of Automotive Engineers' (SAE) standards of automation levels are commonly used in automotive engineering for AVs. Based on conventional AV designs, including perception, decision, and action systems, a theoretical study of the types of actual and possible AV incidents is performed. Furthermore, on-road AV disengagement and accident information are provided publicly are statistically analyzed. The on-road testing results in California indicate that more than 3.7 million miles have been tested for AVs by various manufacturers between 2014 and 2018. The AVs are frequently manually taken over by human operators, and the frequency of disengagement varies greatly between manufacturers, ranging from 2 104 to 3 disengagements per mile. Furthermore, 128 incidents have been reported over 3.7 million miles, with automatic mode accounting for about 63.3 percent of all accidents. A minor amount (approximately 6.3 percent) of overall incidents are directly attributable to AVs, whereas the other parties, including pedestrians, cyclists, motorbikes, and conventional cars, begin 93.7 percent of the accidents. These safety hazards revealed during on-road testing, as indicated by disengagements and real accidents, indicate that most accidents are caused by other road users. This indicates that detecting and avoiding safety risk created by other parties will be essential to make safe decisions to prevent fatal accidents.

References

[1] J. M. Anderson, N. Kalra, K. D. Stanley, P. Sorensen,
C. Samaras, and O. A. Oluwatola, *Autonomous Vehicle
Technology: A Guide for Policymakers*, RAND Corporation,
Santa Monica, CA, USA, 2016.
[2] F. M. Favar`o, N. Nader, S. O. Eurich, M. Tripp, and
N. Varadaraju, "Examining accident reports involving autonomous
vehicles in California," *PLoS One*, vol. 12, 2017.
[3] A. Millard-Ball, "Pedestrians, Autonomous vehicles, and
cities," *Journal of Planning Education and Research*, vol. 38,
no. 1, pp. 6–12, 2018.
[4] A. Rodr´ıguez Palmeiro, S. van der Kint, L. Vissers, H. Farah,
J. C. F. de Winter, and M. Hagenzieker, "Interaction between
pedestrians and automated vehicles: a Wizard of Oz experiment,"
*Transportation Research Part F: Traffic Psychology and
Behaviour*, vol. 58, pp. 1005–1020, 2018.
[5] L. C. Davis, "Optimal merging into a high-speed lane dedicated
to connected autonomous vehicles," *Physica A: Statistical
Mechanics and its Applications*, vol. 555, Article ID
124743, 2020.
[6] R. E. Stern, Y. Chen, M. Churchill et al., "Quantifying air
quality benefits resulting from few autonomous vehicles
stabilizing traffic," *Transportation Research Part D: Transport
and Environment*, vol. 67, pp. 351–365, 2019.
[7] US Department of Transportation National Highway Traffic
Safety Administration, *Critical Reasons for Crashes Investigated
in the National Motor Vehicle Crash Causation Survey*,
NHTSA, Washington, DC, USA, 2015.
[8] V. Nagy and B. Horv´ath, "*e effects of autonomous buses to
vehicle scheduling system," *Procedia Computer Science*,
vol. 170, pp. 235–240, 2020.
[9] M. W. Levin, M. Odell, S. Samarasena, and A. Schwartz, "A
linear program for optimal integration of shared autonomous
vehicles with public transit," *Transportation Research Part C:
Emerging Technologies*, vol. 109, pp. 267–288, 2019.
[10] X. Ge, X. Li, and Y. Wang, "Methodologies for evaluating and
optimizing multimodal human-machine-interface of autonomous
vehicles," in *Proceedings of the SAE Technical Paper
Series*, Detroit, MI, USA, 2018.
[11] L.-J. Tian, J.-B. Sheu, and H.-J. Huang, "*e morning commute
problem with endogenous shared autonomous vehicle
penetration and parking space constraint," *Transportation
Research Part B: Methodological*, vol. 123, pp. 258–278, 2019.
[12] Y.-C. Lee and J. H. Mirman, "Parents' perspectives on using
autonomous vehicles to enhance children's mobility,"
Transportation Research Part C: Emerging Technologies,
vol. 96, pp. 415–431, 2018.
[13] D. J. Fagnant and K. Kockelman, "Preparing a nation for
autonomous vehicles: opportunities, barriers and policy

recommendations," *Transportation Research Part A: Policy and Practice*, vol. 77, pp. 167–181, 2015.

[14] R. Bennett, R. Vijaygopal, and R. Kottasz, "Willingness of people who are blind to accept autonomous vehicles: an empirical investigation," *Transportation Research Part F: Traffic Psychology and Behaviour*, vol. 69, pp. 13–27, 2020.

[15] C. Włodzimierz and G. Iwona, "Autonomous vehicles in urban agglomerations," *Transportation Research Procedia*, vol. 40, pp. 655–662, 2019.

[16] T. Z. Zhang and T. D. Chen, "Smart charging management for shared autonomous electric vehicle fleets: a puget sound case study," *Transportation Research Part D: Transport and Environment*, vol. 78, Article ID 102184, 2020.

[17] L. Zhang, F. Chen, X. Ma, and X. Pan, "Fuel economy in truck platooning: a literature overview and directions for future research," *Journal of Advanced Transportation*, vol. 2020, Article ID 2604012, 10 pages, 2020.

[18] J. Farhan and T. D. Chen, "Impact of ridesharing on operational efficiency of shared autonomous electric vehicle fleet," *Transportation Research Part C: Emerging Technologies*, vol. 93, pp. 310–321, 2018.

[19] M. Lokhandwala and H. Cai, "Siting charging stations for electric vehicle adoption in shared autonomous fleets," *Transportation Research Part D: Transport and Environment*, vol. 80, Article ID 102231, 2020.

[20] I. Overtoom, G. Correia, Y. Huang, and A. Verbraeck, "Assessing the impacts of shared autonomous vehicles on congestion and curb use: a traffic simulation study in the Hague, Netherlands," *International Journal of Transportation Science and Technology*, 2020.

[21] P. Koopman and M. Wagner, "Autonomous vehicle safety: an interdisciplinary challenge," *IEEE Intelligent Transportation Systems Magazine*, vol. 9, pp. 90–96, 2017.

[22] C. Xu, Z. Ding, C. Wang, and Z. Li, "Statistical analysis of the patterns and characteristics of connected and autonomous vehicle involved crashes," *Journal of Safety Research*, vol. 71, pp. 41–47, 2019.

[23] B. N´emeth, Z. Bede, and P. G´asp´ar, "Control strategy for the optimization of mixed traffic flow with autonomous vehicles," *IFAC-PapersOnLine*, vol. 52, no. 8, pp. 227–232, 2019.

[24] D. Phan, A. Bab-Hadiashar, C. Y. Lai et al., "Intelligent energy management system for conventional autonomous vehicles," *Energy*, vol. 191, Article ID 116476, 2020.

[25] F. Chen, M. Song, and X. Ma, "A lateral control scheme of autonomous vehicles considering pavement sustainability," *Journal of Cleaner Production*, vol. 256, Article ID 120669, 2020.

[26] F. Chen, M. Song, X. Ma, and X. Zhu, "Assess the impacts of different autonomous trucks' lateral control modes on asphalt pavement performance," *Transportation Research Part C: Emerging Technologies*, vol. 103, pp. 17–29, 2019.

[27] Beijing Innovation Center for Mobility Intelligent, Beijing Autonomous Vehicle Road Test Report, 2018, http://www. mzone.site/.

[28] K. Bimbraw, "Autonomous cars: past, present and future A review of the developments in the last century, the present scenario and the expected future of autonomous vehicle technology," in *Proceedings of the 12th International Conference on Informatics in Control, Automation and Robotics*, pp. 191–198, Colmar, France, July 2015.

[29] L. M. Hulse, H. Xie, and E. R. Galea, "Perceptions of autonomous vehicles: relationships with road users, risk, gender and age," *Safety Science*, vol. 102, pp. 1–13, 2018.

[30] J. Moody, N. Bailey, and J. Zhao, "Public perceptions of autonomous vehicle safety: an international comparison," *Safety Science*, vol. 121, pp. 634–650, 2020.

[31] J. Lee, D. Lee, Y. Park, S. Lee, and T. Ha, "Autonomous vehicles can be shared, but a feeling of ownership is important: examination of the influential factors for intention to use autonomous vehicles," *Transportation Research Part C: Emerging Technologies*, vol. 107, pp. 411–422, 2019.

[32] G. Mordue, A. Yeung, and F. Wu, "*e looming challenges of regulating high level autonomous vehicles," *Transportation Research Part A: Policy and Practice*, vol. 132, pp. 174–187, 2020.

BEI GRIN MACHT SICH IHR WISSEN BEZAHLT

- Wir veröffentlichen Ihre Hausarbeit,
  Bachelor- und Masterarbeit

- Ihr eigenes eBook und Buch -
  weltweit in allen wichtigen Shops

- Verdienen Sie an jedem Verkauf

Jetzt bei www.GRIN.com hochladen und kostenlos publizieren